FIRST TEC Y

Packaging

Author: **John Williams**
Photography: **Zul Mukhida**

Half the fun of getting things
Is ripping wrappers, cutting string,
Pulling ribbons, reading tags,
Reaching deep in paper bags.
A corner felt, a colour spied,
The different shapes that boxes hide.
But the best thing, is all the time,
Knowing what's inside is mine!

an imprint of Hodder Children's Books

FIRST TECHNOLOGY

Titles in this series

Machines
Tools
Wheels and Cogs
Energy
Toys
Packaging

Series editor: Kathryn Smith
Designer: Loraine Hayes

First published in Great Britain in 1993
by Wayland (Publishers) Ltd
Reprinted in 2001 by Hodder Wayland,
an imprint of Hodder Children's Books

British Library Cataloguing in Publication Data

Williams, John
Packaging – (First Technology Series)
I. Title II. Series
688.8

ISBN 0 7502 3411 3

Typeset by DJS Fotoset Ltd, Burgess Hill, Sussex.
Printed and bound in Hong Kong

Acknowledgements
All the photographs in this book were taken by Zul Mukhida, except for the following: John Caldwell 20, 21; Chapel Studios 17 (right); Hulton Deutsch 18; Eye Ubiquitous 7; Chris Fairclough 17 (left); Tony Stone Worldwide 15. The artwork on page 28 is by John Yates.

WARNING: Some types of packaging, such as glass bottles, aluminium cans and plastic bags, can be dangerous. Keep young children away from plastic bags. Ensure that children are always supervised when collecting and sorting through packaging.

Words printed in **bold** appear in the glossary on page 31.

Coke
DIET
Tango
Coca-Cola

Almost everything we buy comes in a package. Boxes, bottles, bags, cartons, sweet wrappers and tins are all kinds of packaging.

How many different kinds of packaging can you see in this lunch box? What do you think the packets are made of?

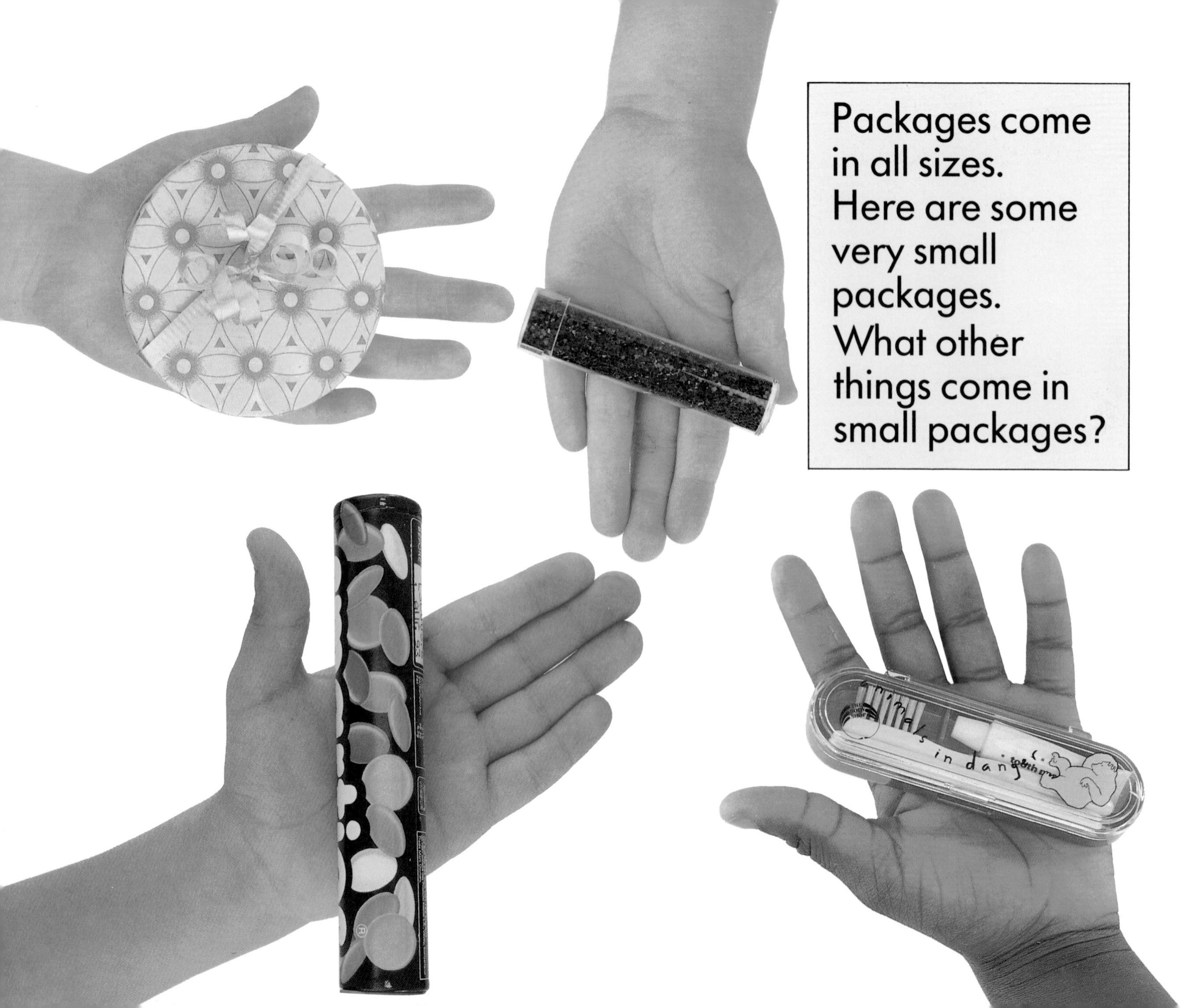

Packages come in all sizes. Here are some very small packages. What other things come in small packages?

Some packages are very big and heavy. Can you think of anything that comes in a large package?

Packages often have words and pictures on them so that we know what is inside.

Look at the pictures on the boxes on the last page. Which toy belongs in which box?

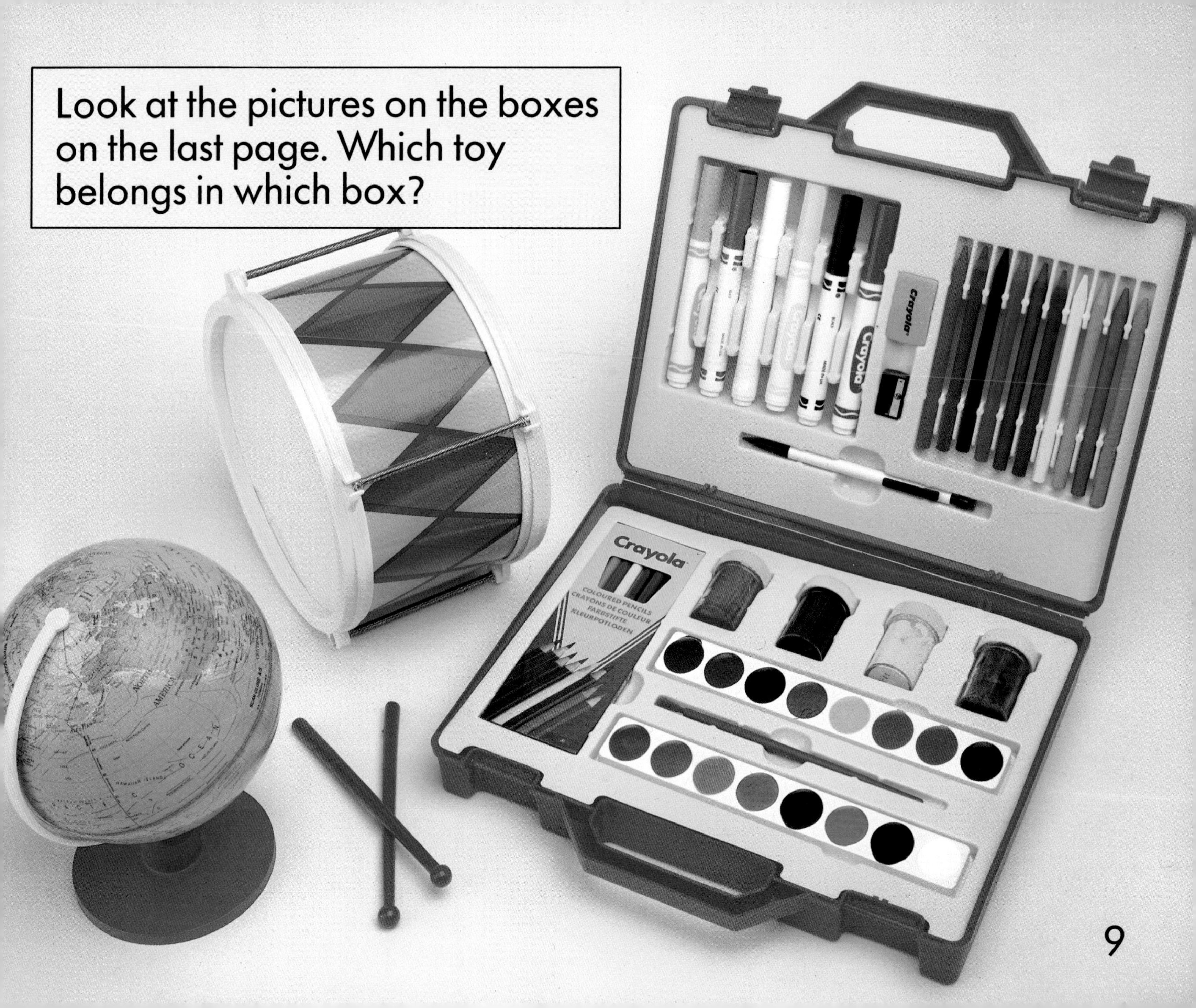

Here are four different kinds of package.

Which of the packages on the last page would you put these things in?

Packages often have special labels printed on them. They tell us more about what is inside. Labels also tell us what to do with the packages when we have finished with them.

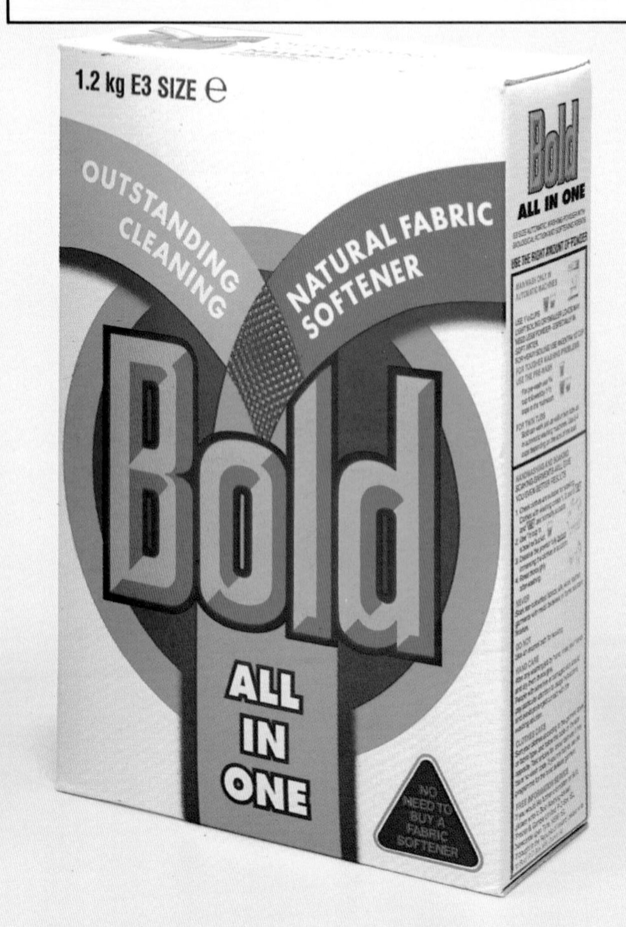

What messages can you see on these packages and tins?

This man is using a trolley to carry lots of packages. Can you think of other ways to carry or move packages?

The large boxes being loaded on to this ship are called **containers**. They are made of metal. They are too large and heavy for a person to move. A crane lifts them on to the ship.

Sometimes we need to send a package by post.

The parcel has to be wrapped in strong paper.

The address must be written on the front.

String is tied round the parcel.

When a parcel is ready, it goes to the post office. At the post office the parcels are sorted.

A postman delivers them to the right address.

This picture shows what a sweet shop looked like eighty years ago. Most of the sweets were kept in jars. In those days sweets were put in brown paper bags.

These days sweets come in different kinds of wrappers.
They are very bright and colourful.

These pictures show how milk is put in bottles.

This machine fills the bottles with milk and puts lids on them.

This machine washes the milk bottles.

This man is loading the bottles into crates.

The crates are stored in a cold place, to keep the milk fresh.

Many packages are made from **aluminium**, **plastic** or glass. We can **recycle** these materials.

Sometimes, when a bottle or a box is empty, we can **re-fill** it. Joseph is re-filling this plastic bottle with washing liquid. Can you think of other ways to use old packages?

Some packages are made to hold special things. This bunch of flowers is wrapped in clear plastic. You can see the beautiful flowers inside.

On special occasions we give presents. These presents are wrapped in brightly coloured paper, so that we cannot see what is inside.

This is a very
special box.
It is a cat box.
Ruskin has
a blanket
in his box to
make him
comfortable.

A suitcase is a special kind of package, too. When we go away, we pack things we need inside it. Can you think of any other types of packaging?

Making a box

You will need some stiff coloured cardboard, a pair of scissors, a sharp pencil, a ruler, a tube of glue and some crayons.

6 cm

A

6 cm

1. Ask a grown-up to draw this shape on card and help you cut it out very carefully. When it is folded, this shape will make the box.

Note to teachers and parents: this diagram is not to scale.

2. Draw some pictures on the box-plan, to show what you would like to put inside the box.

3. Fold the box-plan along the dotted lines, to make the box. Put glue on flaps A and B to stick your box together.

4. Do not stick down flap C. This makes the lid of your box.

NOTES FOR TEACHERS AND PARENTS

Packaging is a major industry. It uses many different materials; plastics, paper, wood, metal and glass are some of the substances used. Much time and money is spent on researching ways to package articles as diverse as cornflakes and crockery. Care is taken to design new packaging, so that it is both eye-catching and informative.

Children should be given the opportunity to explore and experiment with the different materials used in the manufacture of various packages. They will discover that what is suitable for one type of product may not be suitable for another.

Children should be able to sort collections of packages and containers. They should be encouraged to put them into sets, depending on their shape, colour, or material. Cans can either be non-magnetic aluminium or magnetic tin-plate.

Children can also design their own labels. They should be aware of the special information on labels, much of it required by law.

It is difficult for very young children to devise ways to test packages for strength, or wear and tear. However, they can complete the first part of the design process by making their own plans and models. This can involve designing special containers for delicate materials, or living things, or designing their own stamps and labels for parcels.

Making their own box

Children may need some help to draw the net (plan). However, they should be encouraged to draw as much as they can, and to understand that some precision is required with the use of ruler and pencil. Once the net is cut out, the children should colour the box before it is folded into a shape. There is an element of technology in this process, but mathematics and art are also involved.

Teachers should be aware of the opportunity to involve other areas of the curriculum in this work. Apart from those already mentioned, language, history and even geography can be included.

GLOSSARY

Aluminium A kind of metal.

Containers Special boxes that are used to carry things. They are very large and usually made of metal.

Materials What things are made of. Metal, plastic, paper, card and glass are all different types of materials.

Plastic A special kind of material that can be made into any shape we like.

Recycle To use something again.

Re-fill To fill a bottle or box again, after it is empty.

INDEX